POLYMER SCIENCE AND TECHNOLOGY

FATIGUE OF POLYMER MATRIX COMPOSITES AT ELEVATED TEMPERATURES

POLYMER SCIENCE AND TECHNOLOGY

Additional books in this series can be found on Nova's website under the Series tab.

Additional E-books in this series can be found on Nova's website under the E-books tab.

MATERIALS SCIENCE AND TECHNOLOGIES

Additional books in this series can be found on Nova's website under the Series tab.

Additional E-books in this series can be found on Nova's website under the E-books tab.

POLYMER SCIENCE AND TECHNOLOGY

FATIGUE OF POLYMER MATRIX COMPOSITES AT ELEVATED TEMPERATURES

JOHN MONTESANO
ZOUHEIR FAWAZ
KAMRAN BEHDINAN
AND
CHEUNG POON

Nova Science Publishers, Inc.
New York

For permission to use material from this book please contact us:
Telephone 631-231-7269; Fax 631-231-8175
Web Site: http://www.novapublishers.com

NOTICE TO THE READER

Library of Congress Cataloging-in-Publication Data
Fatigue of polymer matrix composites at elevated temperatures / John Montesano ... [et al.].
p. cm.
Includes index.
ISBN 978-1-61761-874-1 (softcover)
1. Polymeric composites--Thermal fatigue. 2. Deformations (Mechanics) I. Montesano, John.
TA418.9.C6F3925 2010
620.1'9204217--dc22
2010033110

Published by Nova Science Publishers, Inc. ✢ New York

CONTENTS

PREFACE

In recent years, advanced composite materials have been frequently selected for aerospace applications due to their light weight and high strength. Polymer matrix composite (PMC) materials have also been increasingly considered for use in elevated temperature applications, such as supersonic vehicle airframes and propulsion system components. A new generation of high glass-transition temperature polymers has enabled this development to materialize. Clearly, there is a requirement to better understand the mechanical behaviour of this class of composite materials in order to achieve widespread acceptance in practical applications. More specifically, an improved understanding of the behaviour of PMC materials when subjected to elevated temperature cyclic loading is warranted.

This book contains a comprehensive review of the experimental and numerical studies conducted on various PMC materials subjected to elevated temperature fatigue loading. The experimental investigations typically focus on observing damage phenomenon and time-dependent material behaviour exhibited during elevated temperature testing, whereas insufficient fatigue test data is found in the literature. This is mainly due to the long-term high temperature limitations of most conventional PMC materials and of the limitations on the experimental apparatus. Moreover, it has been found that few fatigue models have been developed that are suitable for damage progression simulations of PMC materials during elevated temperature fatigue loading. In general fatigue damage simulation models for PMC materials are quite complex due to the nature of the heterogeneous microstructure of these advanced materials. Adding the element of elevated temperature further complicates the physics of the problem due to the probable time-dependent behaviour of the material and the potential effects of aging. Although this review is not exhaustive, the noteworthy results and the trends of the most important studies are presented as well as their apparent shortcomings. Recommendations for future studies are also briefly addressed and the focus of some current research efforts is outlined.

Chapter 1

INTRODUCTION

Considerable progress in the development of composite materials during the past few decades has enabled their widespread utilization for various industrial and recreational applications. In recent years advanced composites have emerged as indispensable materials in the aerospace industry, and as a consequence are more frequently employed due to their high strength-to-weight ratios when compared to conventional metallic components. This is exemplified by considering modern commercial aircraft such as the Airbus A380 and the Boeing 787, both of which utilize composite materials for primary structural load bearing components. The airframe of the A380 that is currently in service is comprised of more than 20% composite materials [1], mainly located in the wingbox interior structure and the aft fuselage section. Once completed, the 787 airframe will have a gross weight that is comprised of approximately 50% composite materials, which results in an aircraft that is 80% composite by volume [2]. The established acceptance of these materials in the aircraft industry and their importance for the future development of more efficient aircraft is apparent.

The integration of composite materials into the propulsion systems of modern commercial and military aircraft has not experienced the same advancement. This is mainly due to the demanding temperature regime that engine components must withstand during standard operation. Nevertheless, there has been some success in using composite materials to manufacture engine components. In the mid-1990's GE successfully integrated polymer matrix composite (PMC) fan blades on the GE90 turbofan engine. Currently GE is developing the next generation turbofan engine GEnx, which comprises of composite fan blades and an entire fan casing manufactured from a braided carbon-fiber PMC material [3]. Since these components are in the cold section of the engine, the ambient operating temperature is typically less-than 100°C. In addition, ceramic-matrix composites (CMC) and metal-matrix composites (MMC) are currently being considered as materials for jet engine components due to their superior heat resistance capabilities. Pratt and Whitney have considered a CMC material for the seals on the exhaust nozzle of the F100 PW 229 military turbine engine, while GE are considering a CMC material for the turbine vanes in the F136 developmental engine. These components are in the hot section of the engine, which can reach temperatures well in excess of 500°C. Clearly, PMC materials would not withstand long-term exposure to this severe temperature environment.

There are however current demands in the industry to manufacture various structural components from composite materials for employment in the moderate temperature regions of jet engines [4], and for next-generation supersonic aircraft fuselage structures [5]. These applications demand long-term exposure to operating temperatures in the 150 - 350°C range. Fiber-reinforced PMC materials with high temperature resins may be suitable candidates for these applications, which will provide weight-saving advantages over conventional metallic components and a reduction in manufacturing costs when compared to MMC and CMC components. A new generation of high glass-transition (T_g) temperature polymers has enabled the current development of high temperature PMC materials. Consequently, high temperature PMC's have been the focus of numerous research efforts over the past 2 decades. Both experimental and numerical studies have attempted to predict and understand the mechanical behaviour and the durability of PMC's at elevated temperatures. More specifically, few fatigue studies on these advanced materials have been presented in the literature. Understanding the fatigue behaviour of advanced composite materials is crucial for predicting their fatigue life and durability. Since aircraft components may be required to survive for over 20 years in service, the accuracy of fatigue life prediction is necessary to ensure the safe-life of composite components.

Chapter 2

FATIGUE BEHAVIOUR OF PMC MATERIALS

Continuous unidirectional, woven or braided fiber-reinforced PMC laminates are commonly used in critical aircraft structural parts. These materials are inhomogeneous and anisotropic, and as such exhibit markedly different behaviour than homogeneous and isotropic materials such as metallic alloys. It is therefore difficult to predict the fatigue properties of these composite materials.

In general, the fatigue behaviour of metallic alloys is well understood and rather predictable. Components made from metals typically exhibit fatigue micro-crack initiation at high stress concentration locations. The gradual growth of these micro-cracks progresses for most of the components lifetime, having little influence on the macroscopic properties of the material. During the final stage, the cracks coalesce to form a larger crack which leads to rapid final failure. Once the visible dominant crack is formed, after a certain number of load cycles, the fatigue life can be determined as long as the initial crack size and its growth behaviour are known. For metallic alloys the macroscopic material properties such as stiffness and strength are unaffected or only slightly affected during fatigue loading, thus simple linearly elastic fracture mechanics models are often adopted to simulate fatigue crack propagation.

Composite components on the other hand exhibit widespread damage throughout the structure without any explicit stress concentrations. Damage can also exist on both microscopic and macroscopic size scales. The common forms of damage (i.e., damage mechanisms) caused by cyclic loading are matrix cracking, fiber fracture, fiber-matrix interface debonding and delamination between adjacent plies [6]. The interaction of these damage mechanisms has been experimentally observed to have a significant influence on the fatigue behaviour [7]. Also since damage commences after only a few loading cycles and progresses upon further cycling, there is typically a gradual stiffness loss in the damaged areas of the material which leads to a continuous redistribution of stress during cyclic loading. As a consequence, simple fracture mechanics-based models are not suitable for composites since the aforementioned damage mechanisms are quite complex and the relationship between stress and strain is no longer linear. In addition, some types of composites such as cross-ply laminates have been found during cyclic loading to reach a state of damage equilibrium, which is deemed a characteristic damage state (CDS) [8]. The progression of matrix cracks in the cross-plies was found to arrest at ply interfaces and at fiber locations, causing the degradation of stiffness to vanish. Therefore, accurate prediction of composite component fatigue behaviour and fatigue life is a complex task.

Experimental characterization of composites is also difficult due to the challenges in inspecting the aforementioned forms of damage and in measuring the continuous degradation of macroscopic material properties. Factors such as the constituent material properties, the fiber structure (unidirectional, woven, or braided), the laminate stacking sequence, the environmental conditions and the loading conditions (maximum stress, loading frequency) among others influence the fatigue behaviour of composites. This results in laborious and costly experimental programs to characterize the material and to generate sufficient fatigue life data. This further limits the approval of newly developed prediction models since validation of a robust model must be done using various experimental test results.

High temperature exposure during cyclic mechanical loading undoubtedly augments the material behaviour and the progression of the aforementioned damage mechanisms due to the potentially complex thermo-mechanical interactions. Additional property degradation mechanisms such as physical and chemical aging may continuously alter the composite properties with time, specifically impacting the polymer matrix behaviour. The influence of the time-dependent material behaviour on the fatigue damage mechanisms will also be significant at severe operating temperatures. This may in fact be the case at temperatures well below the T_g of the polymer matrix [9]. Development of a comprehensive prediction methodology for fatigue behaviour or fatigue life prediction at elevated temperatures is consequently an even more difficult task. Moreover, long-term fatigue testing at elevated temperatures poses additional difficulties due to the severe test environment, which may limit utilization of conventional fatigue testing equipment and techniques. As indicated, the continued development of high temperature polymers and their respective composites has enabled this state-of-the-art research to persist on these advanced materials.

Chapter 3

Development of High Temperature Polymers

For a number of decades now, many researchers have considered the effects of elevated temperature exposure on various polymers. In the 1970's and 1980's, a number of high temperature polymer resins were developed by NASA as part of a larger research effort and considered as potential candidates for fiber-reinforced composite materials. Through this research, two groups of polymers known as linear polyimides and addition aromatic polyimides were developed [10]. The linear polyimides are attractive since they are both tough (i.e., high damage tolerability) and have remarkable thermal stability over a wide temperature range. The addition aromatic polyimides are more brittle, but have highly cross-linked molecular structure [11], which is beneficial for higher temperature stability where linear polyimides may fail. The main setback with these types of polyimides is that they contain known carcinogenic by-products and are very hazardous, which poses many manufacturing difficulties and risks.

The first group of elevated temperature polyimide resins widely produced by NASA for use in fiber-reinforced composites was developed using a polymerization of monomer reactants (PMR) approach [12]. These addition-type polyimides were developed to have excellent thermal stability, ease of manufacturability and the ability to withstand temperatures in excess of 300°C (i.e., a trade-off between linear and addition aromatic polyimides). The static strength of these polyimides over long-term high temperature exposure was found to be fairly stable. The main derivative of this group of polyimide resins to be employed for high temperature aerospace applications is PMR-15. Many studies were conducted by NASA to improve the manufacturability and mechanical performance of PMR-15, and to 'tailor-make' this polyimide resin for use in fiber-reinforced composites [13]-[15]. Experimental studies were later conducted on fiber-reinforced PMR-15 composites [4]. The static mechanical property degradation, weight-loss, coupon dimensional changes, and surface thermal oxidation effects due to long-term aging at elevated temperatures were all considered during testing. Aging temperatures were limited to 350°C. Surface thermal oxidation was believed to be a significant contributor to material property degradation for aging greater-than 100 hours, causing microvoids and microcracks to initiate just below the damaged material surface layer. Although testing has been conducted at higher temperatures for PMR-15 composites, the maximum useful long-term operating temperature is approximately 260°C for jet engine applications.

Due to the successful development and wide regard of PMR-15, additional polyimide resins were subsequently developed for high temperature composite applications. NASA also developed AMB-21 [16] and DMBZ-15 [17] high temperature polymers. These thermoset polyimides have similar properties and temperature capabilities as PMR-15, but without the hazardous carcinogenic compounds. In fact DMBZ-15 has a higher wear resistance and a slightly higher T_g when compared to PMR-15, which makes it suitable for long-term exposure at temperatures >300°C. Moreover, Dupont developed a thermoplastic polyimide Avimid K3B, which has been considered for supersonic transport aircraft [18]. The continuous maximum operating temperature for K3B is approximately 180°C. A number of additional thermoset and thermoplastic polyimide resins such as R1-16, PETI-5 and PIXA among others have also been considered for PMC components on supersonic aircraft with the same temperature limitations [5]. Finally, a number of high temperature BMI polymers have been developed and used in the industry. Common BMI polymers include 5250 and 5260 developed by Cytec Engineered Materials, as well as F655-2 developed by Hexcel Corporation. These polymers have a continuous maximum operating temperature of approximately 150°C.

Although a number of high temperature polymers and their respective fiber-reinforced composites have been developed, there has been little use of these materials in high temperature load bearing applications. Additionally as already indicated, few studies have been conducted that consider the fatigue behaviour of these PMC materials at elevated temperatures.

Chapter 4

REVIEW OF ELEVATED TEMPERATURE FATIGUE STUDIES

This review aims to chronologically delineate the most important accomplished fatigue studies on high temperature PMC materials. First, a discussion of the high temperature experimental work conducted on these advanced materials will be presented. This is followed by a presentation of the subsequently conducted numerical studies.

4.1. EXPERIMENTAL

Most experimental studies focus on temperature-dependent material property degradation during static loading or isothermal aging test conditions. There are few experimental studies that consider fatigue loading at elevated temperatures. These studies will be presented, and the focus of the discussion will be on the indicated observable effects of time and temperature on the fatigue behaviour and corresponding damage mechanisms. The discussion will also include detail of the experimental test protocol and test equipment for specific studies.

Lo et al [19] developed a fiber-reinforced composite which was manufactured with CSPI, a modified polyimide developed at the Chung Shan Institute of Science and Technology, having a T_g of 511°C. Isothermal mechanical fatigue testing of carbon fiber/CSPI and carbon fiber/PMR-15 composites at 450°C was conducted using unidirectional and [0/90/±45] laminates. For a peak fatigue stress level of 60% of the ultimate strength and a loading frequency of 2 Hz, it was found that the fatigue life at room temperature (RT) for the CSPI composite was 10^6 cycles while at 450°C the fatigue life was of the order 10^4 cycles. The material did show a drastic decrease in fatigue life at elevated temperatures, which is not surprising. The CSPI composite demonstrated superior static and fatigue properties at elevated temperatures when compared to the PMR-15 composite. There was however no consideration for tracking the progression of damage or for quantifying material property degradation of the CSPI specimens. In addition, little information is available in the open literature to suggest any application of this material in the industry to date.

Branco et al [20] conducted isothermal fatigue tests at various temperatures, stress ratios and loading frequencies for a glass fiber reinforced unidirectional phenolic resin BPJ 2018L composite up to 200°C. The glass fibers had a surface treatment applied in order to protect them from acid attack. This treatment clearly had an influence on the fiber-matrix bonding characteristics, and thus the fatigue behaviour. Stiffness degradation was monitored for both notched and un-notched specimens using an extensometer. The testing temperature was found to influence the rate of modulus reduction. It was clear that the same specimen subjected to the same loading conditions but at higher temperatures exhibited a consistent stiffness loss, whereas at RT there was little stiffness loss until close to failure. The respective plots of the normalized stiffness versus the normalized number of loading cycles for the phenolic resin composite material are shown in Figure 1. Not surprisingly, fatigue life was found to decrease with increasing temperature. Also, the fatigue life increased slightly as the loading frequency increased at the same test temperature. There is a clear time-dependence in the response of the material, which depends on the rate of stress application. Matrix cracking, fiber-matrix debonding and fiber fracture were all observed in the failed specimens using a SEM post-test. Debonding between the fibers and the matrix was deemed to be the dominant damage mechanism causing fatigue failure. Branco et al [21] later conducted the same tests using composite laminates manufactured with the same phenolic matrix with various stacking sequences. It was found that the manufacturing method (i.e., hand lay-up or pultrusion) strongly influenced the fatigue life of a specimen at elevated temperatures.

Uematsu et al [22] studied delamination behaviour of unidirectional fiber-reinforced PEEK thermoplastic laminates subject to isothermal fatigue loading at 200°C. Double cantilever beam specimens were used to facilitate ply delamination during fatigue. Delamination was initiated artificially using a thin film located between adjacent laminate plies.

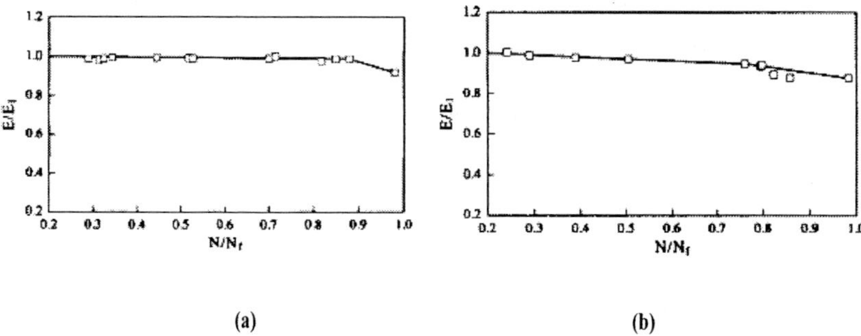

Figure 1. Plots of stiffness degradation at (a) RT and (b) 100°C [20].

Material stiffness was shown to decrease with increasing temperature during static testing. A fracture mechanics-based analysis was used to formulate the change in energy release rate (G_I) and the corresponding stress intensity factor (K). Constant load isothermal creep tests revealed that delamination growth is continuously steady; an initial spike in delamination crack size is due to matrix degradation, while fiber bridging slows the rate of crack growth until final fracture. Fatigue loading at elevated temperatures significantly increases the delamination growth rate, which is highly dependent on the loading frequency. During higher frequency loading the crack propagation rate was completely frequency-dependent or cycle-dependent (i.e., da/dN is proportional to ΔK). At lower loading frequency the crack propagation rate was independent of the frequency and completely time-dependent or creep-dependent (i.e., da/dt is proportional to K). The threshold frequency between the low/high regions was found to be approximately 0.05 Hz, as shown in the plot of crack propagation rate (da/dt) versus the inverse of the loading frequency ($1/v$) in Figure 2. This shows that there is little interaction between creep and fatigue mechanisms, which may seem surprising. Fatigue was therefore classified as being either time-dependent or cycle-dependent.

Sjogren and Asp [23] also conducted a similar study to determine the effects of temperature on delamination growth in prepreg fiber-reinforced epoxy laminates subject to flexural and mixed-mode bending fatigue loading at 100°C. Delamination was also initiated artificially between selected adjacent plies. The effect of temperature on the energy release rate values for delamination growth was similarly found, where critical and threshold energy release rates decreased with an increase in test temperature. Delamination was consequently deemed to be the dominant damage mechanism causing final failure.

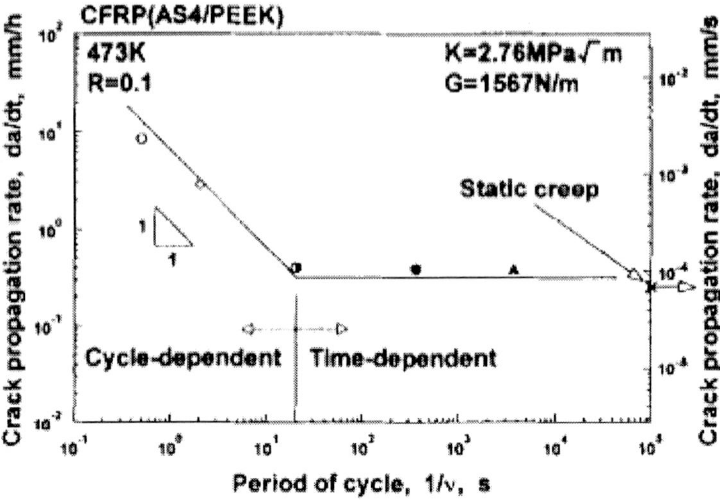

Figure 2. Plot of crack propagation rate as a function of $1/\nu$ [22].

Gyekenyesi et al [24] conducted an experimental study on a woven fiber-reinforced AMB21 polyimide resin matrix composite. Mechanical fatigue loading at test temperatures of 255°C was conducted using a quartz lamp as a heat source, water-cooled hydraulic grips for loading the specimen, and an air-cooled extensometer to measure axial strain. Although the emphasis of the study was on the experimentation methodology, some noteworthy results for the material were obtained. High temperature static tensile tests were initially conducted, and it was found that temperature had little influence on the modulus, the ultimate stress and the strain. Fatigue loading at 255°C proved to have some influence on the fatigue life (N_f) when compared to similar RT tests. It must be noted that the coupons tested showed a large variation in fatigue lives, which was stated to be the result of poor quality specimens that included a number of defects. Ratcheting of the stress-strain curve was however observed during tension-tension fatigue, and was attributed to fiber fracture, matrix cracking, chemical degradation and mass loss due to elevated temperature and viscoelastic deformation. Ratcheting was also observed during RT testing with a similar trend in the stress-strain curve. Successive stress-strain curves for various load cycles are shown in Figure 3 for elevated temperature testing, which clearly illustrates this ratcheting behaviour. Maximum strain was used as the damage metric; there was a significant initial increase in maximum strain, followed by a gradual increase, and ending in a sudden increase before failure. The stiffness of the composite was also monitored and found to decrease continuously.

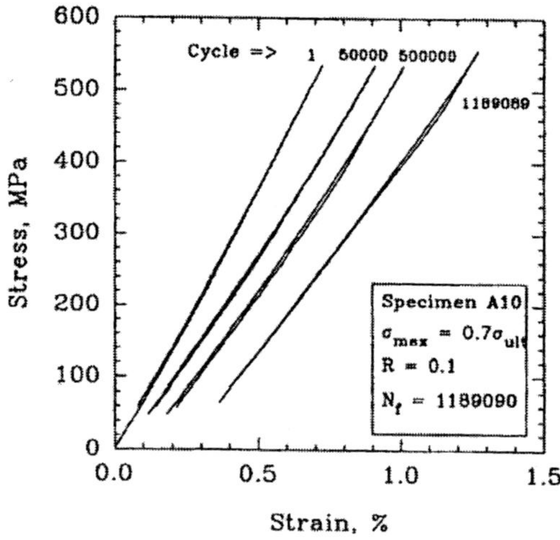

Figure 3. Strain ratcheting at elevated temperature [24].

Miyano et al [9] studied the effects of time and temperature on the flexural behaviour of unidirectional CFRP laminates subjected to fatigue loading. Two resin materials were considered for composite manufacturing: a general purpose epoxy 25C and a high T_g polycyanate resin RS3. Static flexural testing revealed that as the temperature increased, the mode of failure changed from tensile at lower temperatures to compressive at higher temperatures. Micro-buckling of the fibers on the compression side of the flexural specimen was observed at higher temperatures due to matrix softening. This fracture process was also observed during fatigue loading. The fatigue behaviour was remarkably dependent on both temperature and loading frequency (i.e., loading rate or time of exposure), which was attributed to the dominant viscoelastic behaviour of the matrix material. Testing revealed that increasing the test time (i.e., decreasing the loading frequency) or the test temperature caused the fatigue strength to decrease. This is illustrated in the stress-cycle (S-N) plot of Figure 4 for the RS3 resin composite. The peak testing temperature for the flexural fatigue tests was 100°C, while the loading frequency varied from either 0.05 Hz or 5 Hz. The damage due to time of exposure was found to be greater than the damage due to the number of loading cycles, which may seem surprising. Miyano and co-workers [25] continued this study where they considered frequency and temperature effects on the flexural fatigue behaviour of woven CFRP laminates manufactured from a high T_g resin 3601. Similar observations were found with the woven laminates.

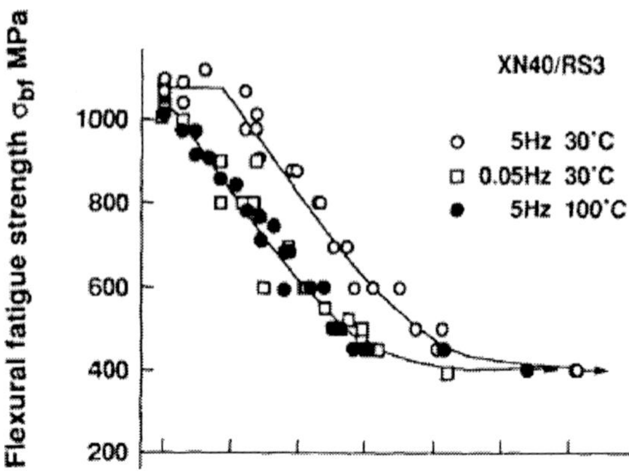

Figure 4. Stress-cycle plot [9].

Case et al [26] studied the behaviour of notched unidirectional-ply fiber-reinforced K3B resin laminates. The specimens were fatigue tested at an elevated test temperature of 177°C and a constant loading frequency of 10 Hz. A convection oven was used to keep the test temperature constant, while strain gages were used for strain measurements. The strain gages did not properly bond to the specimens, thus any strain data was deemed inconclusive in this study. X-ray radiographic images near the specimen notch were also taken at various cycles in order to track damage development. Note that the fatigue test was interrupted in order to access the specimen for x-ray imaging. The dominant damage mechanisms were found to be delamination and matrix cracking. Elevated temperature was found to accelerate the delamination during cycling, while matrix cracking was also more prominent at an elevated test temperature.

Castelli et al [27] tested a chopped fiber-reinforced PMR-15 polyimide matrix composite subject to combined thermo-mechanical fatigue loading at a maximum test temperature of 260°C. The material was considered for a compressor stage component in a jet engine, thus a realistic thermo-mechanical test was conducted. Fatigue damage was tracked macroscopically through deformation and stiffness measurements using an air-cooled extensometer. Image analysis was used to characterize the fiber distribution orientation since the location of damage was dependent on the fiber density. Optical microscopy and SEM were used to characterize local microscopic damage. A quartz lamp system was used for heating the specimens, while an MTS load cell with water-cooled hydraulic grips was used for mechanical cycling. It was found that thermo-mechanical fatigue loading did not considerably degrade the macroscopic material properties such as axial stiffness; stiffness degradation only occurred early in cycling, and was attributed to fiber straightening. Highly localized microscopic damage was however detected at fiber bundle locations including fiber-matrix interface debonding and matrix cracking after 100 hours of cycling. Creep deformation and thus strain accumulation were however found to be significant during thermo-mechanical fatigue cycling, which was further explored through a series of isothermal stress-hold tests. Time-dependent material behaviour was found to occur at temperatures well below the T_g of the polyimide. Aging did not occur in the material after 100 hours of exposure, which was monitored by tracking the value of T_g for the polyimide matrix material.

Kawai et al [28] studied the off-axis behaviour of unidirectional fiber-reinforced polymer composites subject to an elevated test temperature of 100°C and tension-tension fatigue loading. Two matrix resins were considered: PEEK and a thermoplastic polyimide resin PI-SP. The emphasis of the study was placed on the influence of the matrix properties, the temperature and the off-axis angle on fatigue behaviour of an elementary composite ply. A temperature chamber was used along with high-temperature hydraulic grips to load the specimens at a constant frequency of 10 Hz. Failure surfaces were examined using SEM imaging. It was found that as the off axis angle increases, the fatigue strength of the ply decreases which is no surprise. It was also found that the cyclic elastic strain range ($\Delta\varepsilon = \Delta\sigma/E_x$) plotted versus the number of fatigue cycles produced two distinct linear curves, one for the on-axis (0°) loaded plies and one for the off-axis loaded plies. The normalized plot for the polyimide specimens is shown in Figure 5. This illustrates that the fibers are critical for failure for on-axis loading, while the matrix and the fiber-matrix interface is critical for off-axis loading. Also the test temperature was found to have a minimal affect on the fatigue behaviour for the on-axis loaded plies, which was not the case for the off-axis loaded plies. Off-axis plies exhibited a decrease in the fatigue strength at elevated temperatures. The failure mechanisms for on-axis loading were longitudinal matrix cracking propagating to the end tabs, whereas for off-axis loading matrix cracking and fiber-matrix debonding were the dominant mechanisms in the gage section. Also, the fatigue strength was found to change for different matrix resins, which was attributed to the varying matrix ductility at elevated temperatures, and different fiber-matrix bonding strengths. Matrix ductility was observed to be enhanced at higher temperatures, but fiber-matrix bonding was weaker as found in SEM images.

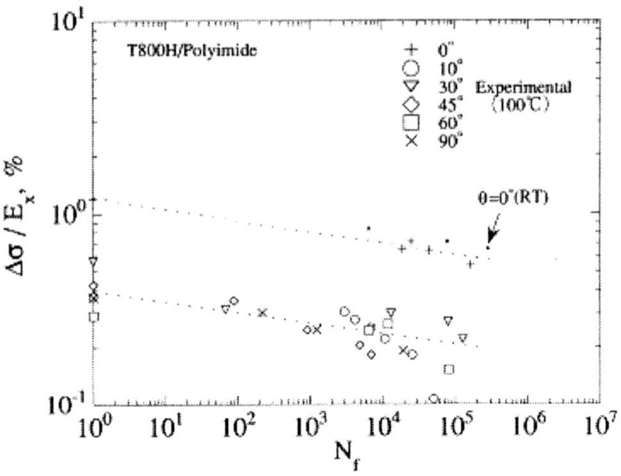

Figure 5. Cyclic strain range vs. number of cycles at 100°C [28].

Counts and Johnson [29] tested the elevated temperature fatigue capabilities of two fiber-reinforced PMC laminates, which were manufactured using PETI-5 and K3B polyimides respectively. These materials were considered for various high temperature applications that required loaded mechanically fastened joints, which in addition to elevated temperature can severely degrade the fatigue performance. Consequently, the focus of the experimental study was on the bolt-bearing capability during fatigue loading at 177°C. The IM7/PETI-5 laminate proved to have superior elevated temperature bearing fatigue properties when compared to the IM7/K3B laminate. It was found that a fatigue endurance limit existed when the maximum cyclic bearing stress was <70% of the ultimate bearing stress. In addition, the bearing fatigue life showed little dependency on the loading frequency (up to 10 Hz) and on pre-fatigue thermal aging (up to 10,000 hours at 177°C). There was no significant time-dependent fatigue behaviour at the elevated test temperature. Increasing the stress ratio (*R*) increased the maximum bearing stress required for failure, which is not surprising. Critical damage mechanisms causing failure were found to be ply delamination, ply buckling and fiber fracture on the compression side of the fastener hole, which were inspected through post-test x-ray radiographic images. Figure 6 includes a local x-ray image of the bearing failure at the fastener location for the IM7/PETI-5 laminate. The aforementioned test variables did not influence the observed damage mechanisms and the bearing failure mode.

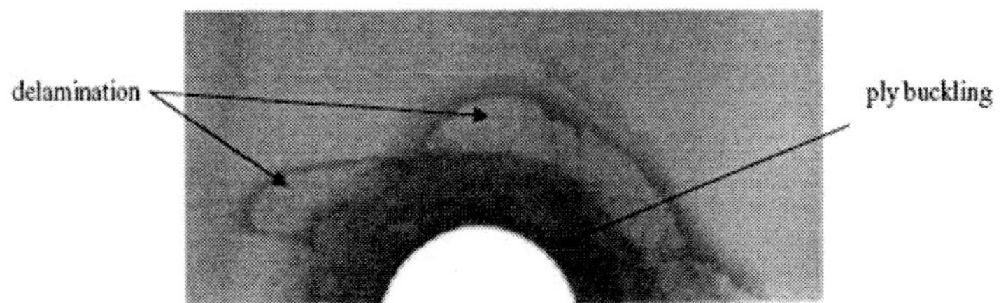

Figure 6. X-ray radiograph of bearing failure of an IM7/PETI-5 specimen [29].

Gregory and Spearing [30] studied the static and fatigue behaviour of a neat resin 977-3 and its unidirectional-ply fiber-reinforced laminate at various temperatures (up to a maximum of 149°C). An experimental methodology was employed whereby the constituent materials were initially tested, and the composite properties were subsequently determined from the constituent material behaviour. In other words, the fracture behaviour of the resin was used to predict the properties of the corresponding composite. The idea was to conduct more constituent material testing in lieu of extensive composite testing since it is simpler and less expensive. Elevated temperature fatigue testing was performed in a custom built temperature chamber furnished with a glass door, which allowed for optical inspection of the composite damage with a traveling microscope. The primary focus was inspecting fatigue delamination growth, which involved use of double cantilever beam composite specimens to facilitate delamination crack growth. The toughness of the resin was shown to be temperature independent up to just below the T_g, where the toughness subsequently decreased. The energy release rate G_{IC} of crack initiation was used as a measure of toughness during tensile testing of notched resin samples. It was also found for the resin that as the loading rate increases, the deformation resistance also increases due to the behaviour of the resin molecular chains. Tensile testing of the composite specimen in flexure revealed that the toughness increased at higher temperatures, which is in contrast to the resin behaviour. The observed fiber bridging effects can be responsible for causing this phenomenon. Delamination of the composite during fatigue cycling proved to be temperature dependent. Delamination was initiated earlier and crack propagation rates were higher at elevated temperatures, which can be attributed to the fact that the resin yield strength decreased with an increase in temperature. Fracture surfaces however were similar for all test temperatures, where brittle fracture was the consistent failure mechanism. This may have been due to the high loading rates as indicated in the work.

The study showed that the toughness and strength of the resin was a key factor in explaining the composite fatigue behaviour.

Shimokawa et al [31] studied the fatigue behaviour of notched woven fiber-reinforced 1053 epoxy laminates at 110°C. Tension-tension (T-T), compression-compression (C-C) and tension-compression (T-C) fatigue tests were all conducted. Extensive matrix cracking was found to have a slight influence on the strength degradation of the material during T-T fatigue testing. The effects of elevated temperature were minimal since the specimens were fiber-dominant in tension. During C-C and T-C fatigue testing, strength degradation was in contrast significant and clearly temperature dependent. This was due to the dependency of the laminate compressive strength on the matrix, which exhibited cracking and elevated test temperatures. Damage progression during T-T fatigue testing began with transverse cracking in the 90° and off-axis tows, followed by tow separation and ply delamination. Note however that the 0° tows were not broken, which explains the minor strength degradation in T-T fatigue. Similar damage was observed during T-C fatigue testing, while less damage was observed during C-C fatigue testing. Note that all damage was observed to occur in the vicinity of the specimen hole using a CCD camera during testing and an X-ray CT scanner post-test. Shimokawa et al [5] observed similar tension and compression fatigue phenomena in a subsequent study in which a fiber-reinforced 5260 BMI resin laminate was fatigue tested at RT and 150°C.

4.2. Prediction Modeling

A number of fatigue models for fiber-reinforced PMC materials have been developed since the 1970's. These models are characteristically empirical, semi-empirical, phenomenological, statistical or mechanistic. The classical empirical and semi-empirical models typically quantify failure by using specific macroscopic failure criteria [6], [32]-[34] and/or by predicting the fatigue life via semi-logarithmic S-N relations [35], [36]. These models do not consider damage or microscopic phenomena, but rely on experimental test data. Phenomenological models often consider the degradation of macroscopically measurable material properties such as the axial strength or stiffness in order to predict the fatigue behaviour. These are typically identified as residual strength [37]-[39] or residual stiffness models [40]-[42] respectively. They are based on the concept of damage accumulation, but do not consider any microscopic effects. Mechanistic models consider the physical interpretation of specific microscopic damage mechanisms such as matrix cracking or ply delamination to predict the progression of damage during cyclic loading [8], [43]-[45]. These models often relate the effects of these damage mechanisms on the degradation of macroscopic material properties. Mechanistic models typically involve finite element (FE) simulation models due to their numerically complex nature. Although purely statistical fatigue models do exist, the statistical probability of physical uncertainties such as constituent material properties or manufacturing voids are typically incorporated into the framework of one of the aforementioned fatigue models. Note that few of these prediction models have been developed with the capability of simulating elevated temperature fatigue loading. Relevant studies are presented here, and the discussion will focus on the manner in which the effects of elevated temperature are introduced into the prediction models.

Shah et al [46] developed a methodology to compute the probabilistic fatigue life of polymer matrix laminates. Matrix degradation effects caused by long-term temperature exposure, thermal fatigue, and mechanical fatigue loading were all accounted for in the simulation model. This was accomplished via a multi-factor interaction equation (MFIE). Specific damage mechanisms are not considered in the model, however various damage modes are considered in the MFIE for degradation of the constituent material properties. The developed material property degradation MFIE is given by:

$$\frac{M_P}{M_{PO}}=\left(\frac{T_{gw}-T}{T_{gd}-T_o}\right)^l\left(1-\frac{\sigma}{S_f}\right)^m\left(1-\frac{\sigma_T}{S_f t_f}\right)^n\left(1-\frac{\sigma_M N_m}{S_f N_{mf}}\right)^p\left(1-\frac{\sigma_T N_T}{S_f N_{Tf}}\right)^q \quad (1)$$

The term M_P is a general material property such as axial stiffness, and M_{PO} is the reference material property. Each parenthesis term represents a particular physical effect that causes degradation of the material property. The five parenthesis terms account for ambient temperature, fiber longitudinal strength, exposure time, mechanical cyclic loading and thermal cyclic loading respectively. The empirical parenthesis exponents factor the respective terms and control the amount of degradation due to that particular term. The time-temperature-stress dependent MFIE is coupled with a statistical model and implemented into a structural multi-scale simulation model to predict the component fatigue life. The statistical model accounts for any uncertainties including void volume ratio, fiber volume ration, constituent material properties and manufacturing errors among others. The multi-scale model simulates the material behaviour of the laminate from the micro-scale constituent level to the component level employing linearly elastic theory. For predicting the laminate strength, ply failure criteria is used. The assumption is that a ply will fail when the maximum stress reaches a critical value, which is assumed to cause the laminate to fail. Finally, the fatigue life is determined using probability density functions which are computed with a fast probability integration technique. This technique is deemed to be more computationally efficient when compared to the commonly employed Monte Carlo technique.

Numerous simulations were conducted to illustrate the capabilities of the simulation model. For simulations with low cyclic load amplitudes, the fatigue life was sensitive to matrix compressive strength, matrix modulus, thermal expansion coefficient and ply thickness. For simulations with high cyclic load amplitudes, the fatigue life was sensitive to the matrix shear strength, fiber modulus, matrix modulus and ply thickness. Viscoelastic material behaviour and progressive damage are not considered in the simulation model, which are seen as major drawbacks. Moreover, the assumption that failure of one ply implies failure of the entire laminate is not necessarily physically accurate. In addition, the MFIE relies heavily on experimental parameters, which may limit the generality and practicality of this model for various types of PMC's.

Miyano et al [25] developed a fatigue strength prediction model for polymeric matrix laminates subjected to flexural cyclic loading at elevated temperatures. The model was based on the aforementioned experimental observations of unidirectional and woven ply laminates. The main assumption of the methodology is that under static, creep and fatigue loading, the failure mechanisms of the laminates are the same (i.e., the compression side of flexure is critical). This led to development of an experimental accelerated testing methodology (ATM) for time-temperature dependent composites. The fatigue strength prediction model [47] is based on a time-temperature superposition principle for an arbitrary viscoelastic matrix composite. Through the ATM, a master plot of exposure time versus ultimate strength was developed for static loading. This plot was created from a number of experimental time/strength curves at various temperatures. Similarly, a master plot of exposure time versus creep strength was also developed for creep loading. Master curves for fatigue strength were also created and drawn on a time (cycle) versus fatigue strength plot. Only one master curve for one stress ratio ($R = 0$) was created and used as a reference for all other stress ratios. The assumption is that there is a linear dependence of the fatigue strength on the stress ratio. All master curves are in fact linear, representing the effects of time and temperature for static, creep and fatigue loading. The equation defining the fatigue strength for an arbitrary loading frequency (*f*), stress ratio (*R*) and temperature (*T*) is:

$$\sigma_f(t_f; f, R, T) = \sigma_{f,R=1}(t_f; f, T)R + \sigma_{f,R=0}(t_f; f, T)(1-R) \quad (2)$$

Here $\sigma_{f,R=1}$ is the creep strength (assumed to be the fatigue strength at $R = 1$), and $\sigma_{f,R=0}$ is the fatigue strength at $R = 0$. The fatigue strength at any time, temperature and number of cycles to failure can be calculated based on this prediction methodology. Only unidirectional and woven ply laminates were considered for analysis, but sufficient correlation to experimental data was shown. The model does however rely on a number of experimental tests for formulation of the master curves which is seen as a limitation. Also, although the effects of viscoelasticity are included in the formulation via the master creep plot, they are not explicitly considered in the model. The assumed linear dependence of fatigue strength on the stress ratio is also seen as a limitation.

Case et al [26] developed a numerical model to predict the fatigue life and residual strength of a unidirectional-ply fiber-reinforced K3B laminate subject to elevated temperature fatigue loading at 177°C. The prediction model employed the critical element method (CEM), which was introduced by Reifsnider [43] a decade earlier. The fundamental concept of the CEM is that the structure is divided into critical and sub-critical elements. For a cross-ply laminate the critical elements would be the 0° plies, while the sub-critical elements would be the 90° plies. The hypothesis is that failure of the critical elements will cause laminate failure, while gradual degradation of the sub-critical elements will result in stress redistribution in the critical elements but not laminate failure. The multi-scale model simulates damage accumulation in the sub-critical elements by gradual degradation of the material stiffness. This is based on micromechanical theory and takes into consideration the constituent material properties, geometry and arrangement, and the progression of damage such as matrix cracking based on empirical data. This subsequently leads to strength degradation in the critical elements, which is based on a homogeneous ply-level strength model and inevitably controls the remaining laminate strength. Appropriate failure criteria, such as maximum stress or strain, are used to determine if failure of the critical plies has taken place. A FE model based on classical laminate theory (CLT) was used to determine the ply-level stresses. The effects of temperature (i.e., viscoelasticity) were not explicitly included in the FE model, but were accounted for in the stiffness degradation of the sub-critical plies by empirical factors. This is seen as a major deficit, which is also stated in the study. In addition, the high temperature data was approximated using RT test data, thus the empirical factors for elevated temperature simulation are clearly inaccurate.

Kawai et al [28] proposed a phenomenological-based damage model for off-axis unidirectional fiber-reinforced polymer lamina subjected to fatigue loading at elevated temperatures. The model relied on a fatigue damage mechanics approach to predict fatigue life. The damage model was based on two internal variables which represented fiber-dominated (ω_f) and matrix-dominated (ω_m) fatigue damage as per experimentally observed phenomena. The growth of these variables is defined by:

$$\frac{\mathrm{d}\omega_f}{\mathrm{d}N} = K_f\left(\sigma^*_{\max}\right)^{n_f}\left(\frac{1}{1-\omega_f}\right)^{k_f} \tag{3a}$$

$$\frac{\mathrm{d}\omega_m}{\mathrm{d}N} = K_m\left(\sigma^*_{\max}\right)^{n_m}\left(\frac{1}{1-\omega_m}\right)^{k_m} \tag{3b}$$

The K, n and k terms are empirical material constants, while $\sigma^*_{\max}$ is the maximum value of the non-dimensional effective stress and is based on the off-axis angle. Using the relations for the two growth variables, the fatigue life for the on-axis and off-axis loaded plies can be derived given the following S-N equations:

$$N_f = \frac{1}{\left(\sigma^*_{\max}\right)^{n_f}} \tag{4a}$$

$$N_f = \frac{1}{\left(\sigma^*_{\max}\right)^{n_m}} \tag{4b}$$

The predicted results were comparable to those obtained using a classical fatigue failure criterion, which was an extension of the well-know Tsai-Hill quadratic criterion (i.e., the constant static strengths are replaced with continuously decreasing functions). Kawai and Maki [48] later incorporated this phenomenological-based ply-level model into a fatigue failure model for cross-ply unidirectional laminates. The prediction model was based on CLT and the in-situ strength of the individual plies. The assumption is that fracture of all the critical on-axis plies is required for laminate failure. They concluded that transverse matrix cracking in the off-axis plies does not affect the fatigue behaviour of the laminate, nor does the elevated temperature or the number of plies. Note that the increase in temperature was obviously found to decreases the fatigue life. They also postulate that fatigue strength is only based on the strength of the individual plies. The model does require extensive experimental parameters to be determined for all desired off-axis angles, and at all desired temperatures and stress ratios. Also, there is no direct consideration for temperature, viscoelastic effects or loading frequency in the fatigue prediction model. Finally, there is no consideration for ply delamination in the prediction model, which may be influential to the fatigue behaviour.

Reifsnider and co-workers [49] developed a model for unidirectional PMC's that relates the longitudinal (E_{11}) and transverse (E_{22}) stiffness to the ambient temperature (T). A modified rule of mixtures is adopted, which includes a temperature-dependent fiber/matrix interface load transfer efficiency factor (η) and the constituent volume fractions (V_f and V_m). The expressions for the longitudinal and transverse modulii in tension are:

$$E_{11}(T) = \frac{E_m(T)V_m + \sum_{i=1}^{N} \lambda_i \exp\left(-\left(\frac{T}{T_i}\right)^{m_i}\right) E_f V_f}{V_m + \sum_{i=1}^{N} \lambda_i \exp\left(-\left(\frac{T}{T_i}\right)^{m_i}\right) V_f} \tag{5}$$

$$E_{22}(T) = \frac{V_m + \sum_{i=1}^{N} \lambda_i \exp\left(-\left(\frac{T}{T_i}\right)^{m_i}\right) V_f}{\sum_{i=1}^{N} \lambda_i \exp\left(-\left(\frac{T}{T_i}\right)^{m_i}\right) \frac{V_f}{E_f} + \frac{V_m}{E_m(T)}} \tag{6}$$

The λ_i term is a relaxation constant for a given material. The fiber modulus E_f is assumed to be constant, while the polymer matrix modulus E_m varies with temperature according to the relation:

$$E_m(T) = E_N \exp\left(-\left(\frac{T}{T_N}\right)^{m_N}\right) + \sum_{i=1}^{N-1} (E_i - E_{i+1}) \exp\left(-\left(\frac{T}{T_i}\right)^{m_i}\right) \tag{7}$$

The integer N is the number of property transitions in the matrix (i.e., glass-transition), T_i and m_i are the transition temperature and statistical coefficient for the ith transition respectively. This formulation allows for the general application of the model for any unidirectional PMC. This model was then implemented into a fatigue prediction model to calculate the end-loaded bending fatigue residual strength for a unidirectional PMC [50]. The fatigue model employs a residual strength integral expression and a stress rupture expression. The residual strength is defined as:

$$Fr = 1 - \int_0^{\tau} (1 - Fa(\tau)) j\tau^{j-1} \mathrm{d}\tau \tag{8}$$

The parameter Fa is the failure function that is based on the applied sinusoidal cyclic load and the loading frequency, j is an empirical parameter, and τ is the characteristic time. For cyclic loading, $\tau = n/N$, where n is the number of cycles and N is the fatigue life. For creep loading, $\tau = t_f$, where t_f is the stress rupture time, which includes the effects of temperature. Failure is said to occur when $Fr = Fa$. The residual strength was determined incrementally in order to combine the effects of creep and fatigue in the prediction model [50]. It was found that the creep effects were less established during fatigue loading when compared to a static rupture case. Also, lower temperature fatigue tests resulted in a higher fatigue life, which implies fatigue dependency. At higher temperatures, significantly lower fatigue lives were predicted, which implies a creep dependent behaviour. Although the model is applicable to all PMC's, the application to more complex fiber geometry other than unidirectional would be difficult. Temperature is considered to calculate the matrix modulus, but viscoelastic behaviour is not explicitly included in the formulation. In addition, the fatigue prediction model does not consider the effects of damage.

Sun et al [51] proposed a fatigue model to predict damage development in polymer matrix unidirectional-ply laminates exposed to high temperatures. A statistical Monte Carlo technique was used to simulate the non-deterministic transverse matrix cracking density progression based on isothermal S-N curves, a damage accumulation model, and a stress analysis model. Miner's rule was employed as the damage accumulation model, and was used to define damage at particular locations in the specimen. The gage length of the specimen was partitioned into 5000 equal elements, and cumulative damage was determined for each element. The damage accumulation model for the *j*th element subject to *K* stages of various stress amplitudes ($S_K{}^j$, $n_K{}^j$) is:

$$D_K^j = \sum_{i=1}^{K} \frac{n_i^j}{N_i^j} \tag{9}$$

A new crack will develop at the *j*th element when $D_K{}^j = 1$. The local stress of the *j*th element is found via a theoretical shear-lag stress analysis model. The remaining life of each element is then determined from the corresponding S-N curve and the damage accumulation equation. The corresponding crack density progression is determined with the Monte Carlo simulation model in Matlab. A statistical probability density function was also proposed for determining transverse crack spacing, which indirectly yielded a correlation to crack density. This function was used in lieu of the Monte Carlo technique to also predict damage progression, which provided consistent results. Only transverse matrix cracking was considered in the analysis up to the characteristic damage state of the material, and only cross-ply laminates were applicable to the study. Also, viscoelastic material effects were not directly considered in the simulation model. Finally, significant experimental testing is necessary to produce the required S-N curves at the required temperatures and to develop the appropriate crack density probability functions.

Jen et al [52] developed a fatigue model which was based on an experimental study of both notched and unnotched unidirectional-ply polymer matrix laminates subject to tension-tension fatigue loading at elevated temperatures (up to 150°C). It is not surprising that at elevated temperatures more rapid degradation of the composite stiffness and strength, and a reduced fatigue resistance was observed. As the temperature was increased closer to T_g, matrix softening initiated fiber/matrix debonding. Subsequently, a modified semi-empirical fatigue model was proposed to predict the durability/life ratio (*S*) for a notched specimen:

$$S = \left[\frac{T_{GW} - T_T}{T_{GO} - T_{TO}} \right]^{1/2} S_O - B\left(\frac{d}{W} \right) \tag{10}$$

The glass-transition temperature at the testing temperature T_T is denoted by T_{GW}, while the glass-transition temperature at room-temperature T_{TO} is denoted by T_{GO}. Also, *B* is an empirical parameter, *d* is the hole diameter and

W is the specimen width. After some manipulation, empirical fitting and setting $S = \sigma$, the final form of the equation relating the applied stress (σ) and the number of cycles (N) can be found for each specific laminate. The model is however too specific, requiring many tests to be conducted for each type of specimen at various temperatures and stress ratios. Also as with other prediction models, viscoelastic effects and damage are not directly accounted for.

Chapter 5

Conclusion

A comprehensive review of the elevated temperature fatigue studies conducted on PMC materials has been presented. Both experimental and numerical studies were considered. The notable results and contributions from each study were discussed, as well as their respective deficiencies. The remaining sections include a discussion on the recommended direction of future work, as well as current research efforts.

5.1. Experimental

As shown, most experimental fatigue studies focus on material property degradation caused by both elevated temperature exposure and mechanical cyclic loading. These studies have generally illustrated the fact that elevated temperature exposure during mechanical cyclic loading induces additional material property degradation mechanisms, which further decreases the effective material stiffness and strength. This in turn has a direct influence on the fatigue behaviour of these composites, causing a decrease in the fatigue life. It is intuitive that at elevated temperatures near the T_g, a polymer matrix will soften and/or decompose consequently affecting the composite material behaviour. This effect may also be amplified in an oxidizing environment during long-term exposure. As an example, consider the PMR-15 polyimide matrix composites developed by NASA. Testing showed that in addition to material property degradation, weight loss and decrease in geometric dimensions during aging occurred, as well as thermal oxidation on the material surfaces [4]. Note that many other studies have shown that these aging effects only occur for long-term exposure (i.e., >100 hours). In addition, testing at elevated temperatures has shown that material creep causes increased ratcheting of the sequential stress-strain curves when compared to equivalent RT tests [24]. Note that at RT ratcheting is typically exhibited by PMC's, but mainly caused by damage accumulation such as matrix cracking. Therefore, time-dependent viscoelastic behaviour and degradation of the matrix material properties likely cause the changes in PMC material properties and an increase in energy dissipation during cyclic loading.

Also, increasing the cyclic loading frequency is found to considerably influence the material behaviour resulting in an increase in the fatigue life of the PMC as indicated in many studies. This may be due to the increase in resistance to deformation caused by the matrix molecular structure and the viscoelastic nature of the material, or in fact be due to the reduction in exposure time at higher loading frequencies. One study reported that at low loading frequencies creep seems to be prevalent having an influence on damage propagation, while at higher loading frequencies damage propagation was more dependent on the number of loading cycles [22]. Although a higher loading frequency has been found to positively influence the fatigue life of PMC's, higher loading frequencies have been shown to cause self-generated heating during isothermal fatigue tests resulting in higher surface temperatures. Bellenger et al [53] studied the changes in surface temperature due to load frequency of a random fiber-reinforced PA66 polyimide matrix composite subject to tension-compression bending fatigue loading at RT. Specimen surface temperatures during high frequency loading were found to increase by more than 100°C, while for lower loading frequencies the surface temperature showed a more moderate increase. Note that short fiber polyimide matrix specimens having low fiber volume fractions were used for this study, which may account for this matrix dominant behaviour. This however could suggest that there may in fact be an upper threshold for loading frequency.

It is also generally observed that the growth rate of specific damage mechanisms increases with increasing temperature during fatigue loading. Shimokawa et al [54] found in their study that matrix cracking facilitated thermal oxidation, which accelerated material property degradation during long-term exposure. Additionally, the damage progression process may alter at elevated temperatures when compared to fatigue testing at room temperature. As an example, fiber-matrix debonding was observed in many studies to occur early on in fatigue testing and is deemed a primary damage mechanism [27], [52]. Debonding is typically followed by cracking of the softened matrix, then delamination and fiber fracture. Conversely, damage progression during fatigue testing at RT has often shown to initiate with matrix cracking or crazing in the off-axis plies [55], [56]. Some studies report that via observable SEM images, the fracture surfaces are similar for specimens fatigue tested at room temperature and at high temperatures [30]. In general regardless of the influence on damage, increased temperature leads to a decrease in the strength of the composite material which consequently decreases the fatigue life.

There are some distinctions in the experimental observations reported which can be attributed to the variations in the type of matrix material (i.e., thermoset or thermoplastic), the composite structure (i.e., unidirectional or woven ply laminates), the loading conditions (i.e., T-T, T-C or C-C; load control or strain control; uniaxial, flexure, or biaxial) and the absolute testing temperature. The mode of loading control is a key factor that will influence the observed phenomena during fatigue testing. The experimental fatigue studies presented are all based on load-control or stress-control testing, whereas strain-control fatigue testing may alter the stress-strain phenomena (i.e., relaxation in lieu of ratcheting) and the progression of damage. Another key factor is the cycling scheme (i.e., T-T, C-C, T-C), which has been shown to reveal contrasting damage mechanisms and failure modes [31]. In either case, the paramount importance of the role of the matrix material on the PMC behaviour during elevated temperature fatigue testing has been demonstrated.

A major shortcoming of all the experimental studies is that there was no attempt to track damage throughout the test specimen continuously and in-situ during the high temperature tests. This is critical for accurate characterization of these materials and for improving the input to the developed prediction models. An improved experimental protocol for high temperature laboratory testing is thus required, specifically continually measuring the strain and tracking damage progression without removing the test specimen from the high temperature environment. Conventional damage monitoring techniques such as x-ray radiography, ultrasound and light microscopy require removal of the test specimen from the loading grips, while other conventional methods are not suitable for elevated temperature applications. A method using a traveling microscope was proposed by Gregory and Spearing [30] to detect delamination in double cantilevered beam specimens; the double cantilevered specimen provided an obvious delamination zone. The capability of this method to detecting other forms of damage within the test specimen must be verified by additional testing.

5.2. Prediction Modeling

As shown, there have been few studies on developing prediction methodologies for PMC's. Although further studies can be found in the literature, they are either continuations of the presented studies or have employed similar models in their work. Of the existing fatigue models that have been presented there are few mechanistic or physically-based progressive damage models, which is seen as a clear gap in the literature. The mechanistic models that have been developed are based on unrealistic assumptions, and/or do not consider all or any forms of observable damage in the simulations. Also, few of the prediction models explicitly consider the viscoelastic effects of the PMC material, adopting linear elastic models or empirical viscoelastic factors in the prediction scheme. Since the time-dependent viscoelastic matrix has a significant influence on the fatigue behaviour, explicit consideration is considered crucial. Moreover, most models rely on extensive experimental testing to extract required empirical factors for the respective formulations. This is seen as another major drawback since the cost of testing is prohibitive in today's aircraft industry where affordability is a major concern. In addition, all models are very specific to particular laminate composites with unidirectional or woven-plies. This limits the robustness of the prediction methodologies.

The use of PMC's for elevated temperature applications such as propulsion system components and supersonic aircraft airframes will undoubtedly increase during the upcoming years due to inevitable environmental and economic demands. Very few PMC's that have been proven to be capable of withstanding long-term exposure at temperatures in the 150 - 350°C range are able to withstand mechanical cyclic loading. Clearly additional studies are required in order to gain confidence in these advanced materials, and to expand their practical usage. The development of accurate and cost-effective fatigue life prediction methodologies for PMC's requires physically-based modeling of damage evolution, as was emphasized by Talreja [57] among others. These models must account for microscopic phenomena such as damage mechanism interactions and manufacturing defects, as well as high temperature effects such as aging, thermal oxidation, matrix degradation and viscoelastic behaviour. It is also in the opinion of the authors that accurate fatigue prediction models must account for physically-based microscopic phenomena and the associated progression of damage. The essential goal is then to relate this microscopic behaviour to the observed macroscopic behaviour of the material. This allows for simulation of the complete path of damage states during cyclic loading, which is essential for appraising the intermediate state of a material or predicting the final state of the material. Although the complexity in developing a mechanistic prediction model may be somewhat high, a few insightful simplifications may be necessary to allow for its use as a practical design tool in the industry without significantly compromising accuracy.

5.3. Current Research

Although accurate prediction models are currently lacking in the open literature, a number of recent studies have been conducted that may provide useful insight on this challenging topic. A current study by the authors [58] is attempting to develop an experimental test methodology for elevated temperature fatigue testing of PMC's by adopting fiber optic sensors for strain and damage detection. Conventional strain monitoring devices such as strain gages and extensometers have their limitations during high-strain cyclic loading. Although strain gages have high static strain ratings, they are not capable of accurately operating at larger strains for many loading cycles [26]. Extensometers have also been shown to slip during high-strain fatigue tests, causing inaccurate strain measurements [24]. Fiber optic sensors have been proven to be sufficient for high-strain cyclic loading at elevated temperatures, which has been supported by another study conducted at room temperature [59]. For damage detection, small-scale optical sensors have been used to detect various forms of damage such as matrix cracking and ply delamination [60], [61]. The multiplexing capabilities of modern high-frequency optical interrogation devices enables continuous monitoring of test specimen damage states during cyclic loading using an array of fiber optic sensors. This state-of-the-art technology shows significant promise to be employed for real-time damage detection during high temperature cyclic loading.

Regarding fatigue prediction modeling, recent studies have provided some insight on the challenges in developing accurate physically-based progressive damage models. Allen and Searcy [62] proposed a multi-scale prediction model for damage in viscoelastic solids. Multi-scale prediction models of composite materials have traditionally accounted for micro-scale phenomena by employing a physically-based local representation of the constituent materials and of the local damage. The analysis results from the local scale are then input into the homogenized global scale model, which is based on the concepts of continuum mechanics. The notion is that physical phenomenon occurring at the local scale can determine the macroscopic behaviour of the material in this hierarchical domain. This is particularly attractive for engineers since FE analysis models are easily incorporated into this modeling methodology. The work by Allen and Searcy [62] extended this notion for viscoelastic composite materials. Talreja [63] developed an alternate multi-scale modeling methodology known as synergistic damage mechanics (SDM), which combines the framework of continuum damage mechanics and micromechanics formulations. Although high temperature fatigue simulations were not considered in these studies, the concepts of the design methodologies may in fact be adopted. With the increase in modern computational power, the development of accurate multi-scale models may lead to adequate tools for predicting high temperature fatigue behaviour of PMC's.

ACKNOWLEDGMENTS

The authors would like to thank the Natural Sciences and Engineering Research Council (NSERC) of Canada for a CRD grant in support of this research. The authors are also indebted to the Consortium for Research and Innovation in Aerospace in Quebec (CRIAQ) for launching and sponsoring a greater research endeavour of which this review is a component. The first author greatly acknowledges additional funding in the form of a Canadian Graduate Scholarship (CGS) by NSERC.

REFERENCES

[1] Marsh, G. *Reinf. Plastics* 2002, 46, 40-43.

[2] Hawk, J. "The Boeing 787 Dreamliner: More Than an Airplane". Presented at the *AIAA/AAAF Aircraft Noise and Emissions Reduction Symposium*, Monterey, CA, May 24-26, 2005.

[3] Marsh, G. *Reinf. Plastics* 2006, 50, 26-29.

[4] Bowles, K.J.; Tsuji, L.; Kamvouris, J.; Roberts, G.D. *NASA Technical Report* 2003, TM-211870.

[5] Shimokawa, T.; Kakuta, Y.; Hamaguchi, Y.; Aiyama, T. *J. Comp Mater.* 2008, 42, 655-679.

[6] Jen, M.R.; Lee, C.H. *Int. J. Fatigue* 1998, 20, 617-629.

[7] Harris, B. In *Fatigue in Composites*; Harris, B.; Ed.; Woodhead Publishing Ltd.: Cambridge, England, 2003; pp 3-35.

[8] Highsmith, A.L.; Reifsnider, K.L. In *Damage in Composite Materials*; Reifsnider, K.L.; Ed.; ASTM STP 775; ASTM International: Philadelphia, PA, 1982; pp 103-117.

[9] Miyano, Y.; McMurray, M.K.; Kitade, N.; Nakada, M.; Mohri, M. *Composites* 1995, 26, 713-717.

[10] St. Clair, A.K.; St. Clair, T.L. *NASA Technical Report* 1981, TM-83141.

[11] Meador, M.A.B. *NASA Technical Report* 1987, TM-89838.

[12] Serafini, T.T.; Vannucci, R.D. *NASA Technical Report* 1975, TM X-71616.

[13] Vannucci, R.D. *NASA Technical Report* 1982, TM-82951.

[14] Vannucci, R.D. *NASA Technical Report* 1987, TM-88942.

[15] Vannucci, R.D.; Malarik, D.C. *NASA Technical Report* 1990, LEW-14923.

[16] Tiano, T.; Hurley, W.; Roylance, M.; Landrau, N.; Kovar, R.F. "Reactive Plasticizers for Resin Transfer Molding of High Temperature PMR Composites". Presented at the *32nd SAMPE ISTC*, Boston, MA, November 5-9, 2000.

[17] Xie, W.; Pan, W.; Chuang, K.C. *Thermo. Act.* 2001, 367-368, 143-153.

[18] Sacks, S.; Johnson, W.S. *J. of Thermoplastic Comp. Mater.* 1998, 11, 429-442.

[19] Lo, Y.J.; Liu, C.H.; Hwang, D.G.; Chang, J.F.; Chen, J.C.; Chen, W.Y.; Hsu, S.E. In *High Temperature and Environmental Effects on Polymeric Composites*; Harris, C.E.; Gates, T.S.; Ed.; ASTM STP 1174; ASTM International: Philadelphia, PA, 1993; pp 66-77.

[20] Branco, C.M.; Eichler, K.; Ferreira, J.M. *Theor. App. Fract. Mech.* 1994, 20, 75-84.

[21] Branco, C.M.; Ferreira, J.M.; Fael, P.; Richardson, M.O.W. *Int. J. Fatigue* 1995, 18, 255-263.

[22] Uematsu, Y.; Kitamura, T.; Ohtani, R. *Comp. Sci. Tech.* 1995, 53, 333-341.

[23] Sjogren, A.; Asp, L.E. *Int. J. Fatigue* 2002, 24, 179-184.

[24] Gyekenyesi, A.L.; Castelli, M.G.; Ellis, J.R.; Burke, C.S. *NASA Technical Report* 1995, TM-106927.

[25] Miyano, Y.; Nakada, M.; McMurray, M.K.; Muki, R. *J. Comp. Mater.* 1997, 31, 619-638.

[26] Case, S.W.; Plunkett, R.B.; Reifsnider, K.L. In *High Temperature and Environmental Effects on Polymeric Composites*; Gates, T.S.; Zureick, A.H.; Ed.; ASTM STP 1302; ASTM International: Philadelphia, PA, 1997; Vol. 2, pp 35-49.

[27] Castelli, M.G.; Sutter, J.K.; Benson, D. "Thermomechanical Fatigue Durability of T650-35/PMR-15 Sheet Molding Compound". *ASTM Symposium on Time-Dependent and Non-linear Effects in Polymers and Composites*, Atlanta, GA, May 4-5, 1998.

[28] Kawai, M.; Yajima, S.; Hachinohe, A.; Kawase, Y. *Comp. Sci. Tech.* 2001, 61, 1285-1302.

[29] Counts, W.A.; Johnson, W.S. *Int. J. Fatigue* 2002, 24, 197-204.

[30] Gregory, J.R.; Spearing, S.M. *Comp. Part A* 2005, 36, 665-674.

[31] Shimokawa, T.; Kakuta, Y.; Saeki, D.; Kogo, Y. *J. Comp. Mater.* 2007, 41, 2245-2265.

[32] Hashin, Z.; Rotem, A. *J. Comp. Mater.* 1973, 7, 448-464.

[33] Reifsnider, K.L.; Gao, Z. *Int. J. Fatigue* 1991, 13, 149-156.

[34] Philippidis, T.P.; Vassilopoulos, A.P. *J. Comp. Mater.* 1999, 33, 1578-1599.

[35] Fawaz, Z.; Ellyin, F. *J. Comp. Mater.* 1994, 28, 1432-1451.

[36] Bond, I.P. *Comp. Part A* 1999, 30, 961-970.

[37] Hahn, H.T.; Kim, R.Y. *J. Comp. Mater.* 1975, 9, 297-311.

[38] Hashin, Z. *Comp. Sci. Tech.* 1985, 23, 1-19.

[39] Whitworth, H.A. *Comp. Struct.* 2000, 48, 261-264.

[40] Hwang, W.; Han, K.S. *J. Comp. Mater.* 1986, 20, 154-165.

[41] Whitworth, H.A. *J. Comp. Mater.* 1987, 21, 362-372.

[42] Yang, J.N.; Jones, D.L.; Yang, S.H.; Meskini, A. *J. Comp. Mater.* 1990, 24, 753-769.

[43] Reifsnider, K.L. *Eng. Fract. Mech.* 1986, 25, 739-749.

[44] Allen, D.H.; Harris, C.E.; Groves, S.E. *Int. J. Sol. Struct.* 1987, 23, 1301-1318.

[45] Shokrieh, M.M.; Lessard, L.B. *Int. J. Fatigue* 1997, 19, 201-207.

[46] Shah, A.R.; Murthy, P.L.N.; Chamis, C.C. "Effect of Cyclic Thermo-Mechanical Loads on Fatigue Reliability in Polymer Matrix Composites". Proceedings of the *36th AIAA/ASME/AHS/ASC Structures, Structural Dynamics and Materials Conference*, New Orleans, LA, April 10-13, 1995. Paper No. AIAA-95-1358.

[47] Miyano, Y.; Nakada, M.; Sekine, N. *Comp. Part B* 2004, 35, 497-502.

[48] Kawai, M.; Maki, N. *Int. J. Fatigue* 2006, 28, 1297-1306.

[49] Mahieux, C.A.; Reifsnider, K.L.; Case, S.W. *App. Comp. Mater.* 2001, 8, 217-234.

[50] Mahieux, C.A.; Reifsnider, K.L.; Jackson, J.J. *App. Comp. Mater.* 2001, 8, 249-261.

[51] Sun, Z.; Daniel, I.M.; Luo, J.J. *Mater. Sci. Eng. A* 2003, 361, 302-311.

[52] Jen, M.H.R.; Tseng, Y.C.; Lin, W.H. *Int. J. Fatigue* 2006, 28, 901-909.

[53] Bellenger, V.; Tcharkhtchi, A.; Castaing, P. *Int. J. Fatigue* 2006, 28, 1348-1352.

[54] Shimokawa, T.; Katoh, H.; Hamaguchi, Y.; Sanbonji, S.; Mizuno, H.; Nakamura, H.; Asagumo, R.; Tamura, H. *J. Comp. Mater.* 2002, 36, 885-895.

[55] O'Brien, T.K.; Reifsnider, K.L. *J. Comp. Mater.* 1981, 15, 55-70.

[56] Razvan, A.; Reifsnider, K.L. *Theor. App. Fract. Mech.* 1991, 16, 81-89.

[57] Talreja, R. "Fatigue Damage Evolution in Composites - A New Way Forward in Modeling". Proceedings of the *2nd International*

Conference on Fatigue of Composites, Williamsburg, VA, 4-7 June, 2000.

[58] Montesano, J.; Selezneva, M.; Fawaz, Z.; Behdinan, K.; Poon, C. "Strain and Damage Monitoring of Polymer Matrix Composite Materials at Elevated Temperatures Using Fiber Optic Sensors", Proceedings of the *SAMPE Conference and Exhibition*, Seattle, WA, May 17-20, 2010.

[59] DeBaere, I.; Luyckx, G.; Voet, E.; VanPaepegem, W.; Degrieck, J. *Opt. Lasers Eng.* 2009, 47, 403-411.

[60] Takeda, S.; Okabe, Y.; Takeda, N. *Comp. Part A* 2002, 33, 971-980.

[61] Yashiro, S.; Okabe, Y.; Takeda, N. *Comp. Sci. Tech.* 2007, 67, 286-295.

[62] Allen, D.H.; Searcy, C.R. *J. of Mater. Sci.* 2006, 41, 6510-6519.

[63] Talreja, R *J. of Mater. Sci.* 2006, 41, 6800-6812.

INDEX

O

P

Q

R

S

T

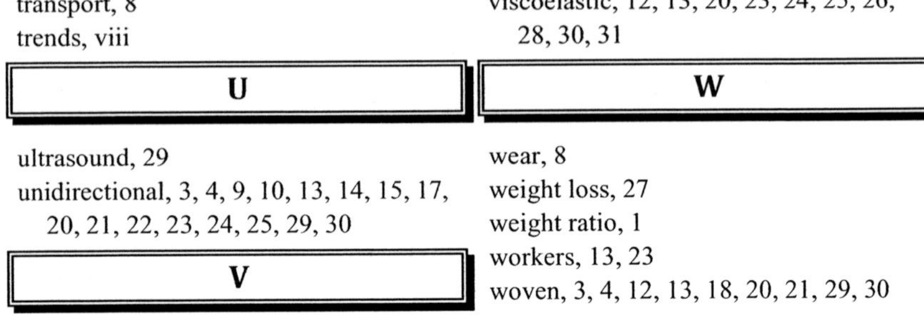